STUDENT GUIDE

GULLIVER'S WORLDS

MEASURING AND SCALING

MathScape™ SEEING AND THINKING MATHEMATICALLY

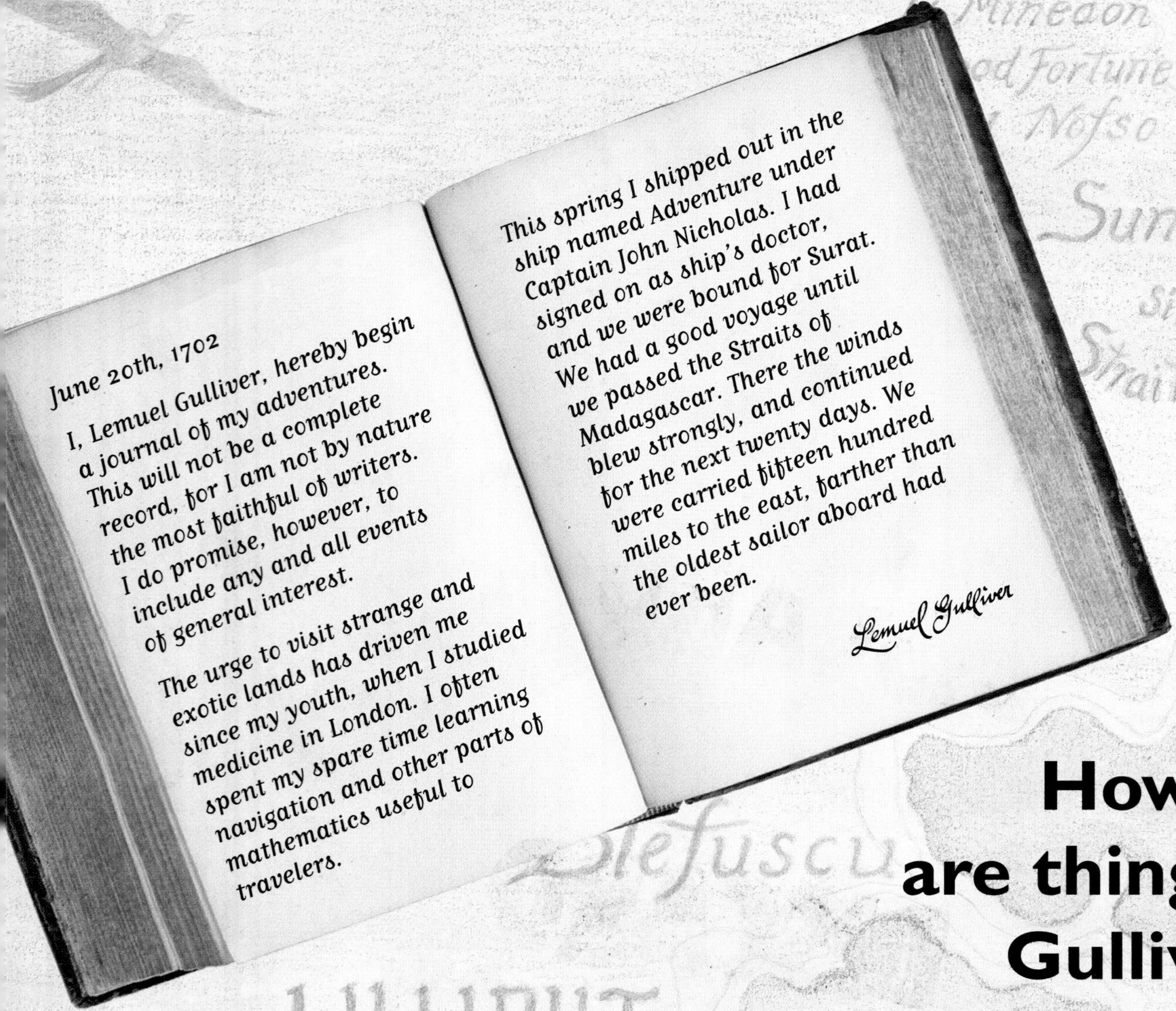

June 20th, 1702

I, Lemuel Gulliver, hereby begin a journal of my adventures. This will not be a complete record, for I am not by nature the most faithful of writers. I do promise, however, to include any and all events of general interest.

The urge to visit strange and exotic lands has driven me since my youth, when I studied medicine in London. I often spent my spare time learning navigation and other parts of mathematics useful to travelers.

This spring I shipped out in the ship named Adventure under Captain John Nicholas. I had signed on as ship's doctor, and we were bound for Surat. We had a good voyage until we passed the Straits of Madagascar. There the winds blew strongly, and continued for the next twenty days. We were carried fifteen hundred miles to the east, farther than the oldest sailor aboard had ever been.

Lemuel Gulliver

How big are things in Gulliver's Worlds?

GULLIVER'S WORLDS

PHASE**ONE**
Brobdingnag

Gulliver's journal holds clues to sizes of things in Brobdingnag, a land of giants. Using these clues, you will find ways to predict the sizes of other things. Then you will use math to create a life-size drawing of a giant object. You will also compare sizes in the two lands. Finally, you will use what you know about scale to write a story set in Brobdingnag.

PHASE**TWO**
Lilliput

Lilliput is a land of tiny people. Gulliver's journal and drawings will help you find out about the sizes of things in Lilliput. You will compare the measurement system in Lilliput to ours. Then you will explore area and volume as you figure out how many Lilliputian objects are needed to feed and house Gulliver. Finally, you will write a story set in Lilliput.

PHASE**THREE**
Lands of the Large and Lands of the Little

Clues from pictures will help you write a scale factor that relates the sizes of things in different lands to the sizes of things in Ourland. You will continue to explore length, area, and volume, and see how these measures change as the scale changes. Finally, you will put together all you have learned to create a museum exhibit about one of these lands.

PHASE ONE

August 29, 1702

We finally sighted land again today. We went ashore near a small creek. I was gone only a short time. Yet when I headed back toward the landing site, the sailors were already rowing frantically out to sea. I could see a huge creature chasing them through the water. It stopped, though, at a sharp reef, and so the sailors escaped.

This was, I admit, of small comfort to me, because I was now alone. Fearing for my safety, I scampered inland. Beyond a steep hill, I discovered tall stalks, about eighteen feet high. They appeared to be wheat. I reached a stone stairway, but finding each step to rise six feet, I was unable to climb it. The trees along its edge were so tall I could not guess their height.

Lemuel Gulliver

Imagine a world in which everything is so large that you would be as small as a mouse. How can you predict how large things will be in this land?

In this phase you will learn to figure out a scale factor that describes how sizes of things are related. You will use the scale factor to create life-size drawings, solve problems, and write stories.

Brobdingnag

WHAT'S THE MATH?

Investigations in this section focus on:

DATA COLLECTION

- Gathering information from a story
- Organizing data to find patterns

MEASUREMENT and ESTIMATION

- Measuring with inches, feet, and fractions of inches
- Estimating the sizes of large objects

SCALE and PROPORTION

- Finding the scale factor that describes the relationship between sizes
- Applying the scale factor to predict sizes of objects
- Creating scale drawings
- Exploring the effect of rescaling on area and volume

1 The Sizes of Things in Brobdingnag

DETERMINING THE SCALE FACTOR

How well can you picture in your mind the events described in Gulliver's journal entry? Here you will gather clues from the journal entry about the sizes of things in Brobdingnag. As you compare sizes of things in Brobdingnag to sizes in Ourland, you will learn about scale.

August 29, 1702

I had not a moment to rest, as another monster was approaching. I now saw that in form he resembled a human being. It was his size—as tall as a ship's mast—that made him appear to be a monster. Scared and confused, I backed away, tripping over an apple core that lay like a log behind me. As I stood up again, the giant began cutting wheat with a great scythe. With every stride he traveled about ten yards closer to me, and I was faced with either being trampled on or cut in two. Therefore, I gave up my hiding place and shouted for his attention.

Compare Sizes to Determine a Scale Factor

How are sizes of things in Brobdingnag related to sizes in Ourland?

A scale factor is a ratio that tells how the sizes of things are related. For example, some model trains use a 20:1 scale factor. This means that each part on the real train is 20 times as large as the same part on the model train. Follow these steps to find the scale factor that relates sizes in Brobdingnag to sizes in Ourland.

1. Make a chart with three columns. Column 1 is for the name of each object. Column 2 is for the size of the object in Brobdingnag. Column 3 is for the size of the corresponding object in Ourland.
2. Fill out column 1 and column 2 with clues you found in the story about the sizes of objects in Brobdingnag. Measure or estimate how big each of the objects would be in Ourland. Enter that information in column 3.
3. Use the information in your chart to figure out a scale factor that tells how sizes of things in Brobdingnag are related to sizes in Ourland.

Object	Brobdingnag	Ourland
Stalk of wheat	About 18 feet	

How big would an Ourland object be in Brobdingnag?

2 A Life-Size Object in Brobdingnag

RESCALING THE SIZES OF OBJECTS

The story continues as Gulliver describes more events from his life in Brobdingnag. In the last lesson, you figured out how sizes in Brobdingnag relate to sizes in Ourland. In this lesson you will use what you know to create a life-size drawing of a Brobdingnag object.

August 30, 1702

I soon became like one more member of the giant's family. In an effort to gain some much-needed privacy, I leaned two large envelopes against a stack of books that measured 10 feet high. Crawling under the envelopes, I had a spacious shelter from prying eyes and a measure of peace. I then made a roomy bed of a small diary with an abandoned eraser as my pillow. Looking around my improvised room, I quickly did a calculation in my head and determined a piece of note paper would be the ideal covering for my long awaited nap. Unfortunately, at this great size the paper was like cardboard and left me longing for my wool blanket back home.

Make a Life-Size Drawing

Choose an object from Ourland that is small enough to fit in your pocket or in your hand. What would be the size of your object in Brobdingnag? Make a life-size drawing of the Brobdingnag object.

1. Make the size of your drawing as accurate as possible. Label the measurements.
2. After you finish your drawing, figure out a way to check that your drawing and measurements are accurate.
3. Write a short description about how you determined the size of your drawing and how you checked that the drawing was accurate.

How can you figure out the size of a Brobdingnag object?

Investigate the Effect of Rescaling on Area

How many of the Ourland objects does it take to cover all of the Brobdingnag object? Use your drawing and the original Ourland object to investigate this question. Write about how you figured out how many Ourland objects it took to cover the Brodingnag object.

3 How Big Is "Little" Glumdalclitch?

ESTIMATING LENGTHS, AREAS, AND VOLUMES

What if Glumdalclitch visited Ourland? Would she fit in your classroom? Drawing a picture of "little" Glumdalclitch in actual Ourland measurements would take a great deal of paper. To get a sense of the size of a very large object, it is sometimes easier to use estimation.

Use Estimation to Solve Problems

Estimate the size of each Brobdingnag object. Answer the questions about how the size of the Brobdingnag object compares to size of the same Ourland object and explain how you found each answer.

1. Could a mattress that would fit Glumdalclitch fit in the classroom? How much of the floor would it cover? How many Ourland mattresses would it take to cover the same amount of floor?
2. How big would Glumdalclitch's notebook be? How many sheets of our notebook paper would we need to tape together to make one sheet for her notebook?
3. How big a shoe box do you think Glumdalclitch might have? How many of our shoe boxes would fit inside hers?
4. How many slices of our bread would it take to make one slice of bread big enough for Glumdalclitch to eat?

Why are so many Ourland objects needed to cover or to fill a Brobdingnag object?

How do objects from Ourland and Brobdingnag compare in length, area, and volume?

BROBDINGNAG
Discovered A.D. 1703

4 Telling Tales in Brobdingnag

USING ACCURATE SIZES IN CREATIVE STORIES

Imagine how it would be for you to visit Brobdingnag. By now you have a good understanding of the scale factor in Brobdingnag. You can use what you know to write your own story. You will see that good mathematical thinking is important in writing a believable story.

Write a Story Using Accurate Dimensions

Choose one place in Brobdingnag. Imagine what it would be like to visit that place. Describe in detail the place and at least one adventure that happened to you there.

1. Write a story about Brobdingnag. Make the story believable by using accurate measurements for the objects you describe.
2. Include a size description of at least three objects found in that place.
3. Write a believable title. The title should include at least one size comparison between Brobdingnag and Ourland.
4. Record and check all of your measurements.

How can you use rescaling to write a story about Brobdingnag?

Summarize the Math Used in the Story

After you write your story, summarize how you used math to figure out the sizes of things in the story. Include the following in your summary:

- Make a table, list, or drawing showing the sizes of the three objects in both Ourland and Brobdingnag.
- Explain how you used scale, estimation, and measurement to figure out the sizes of these objects.

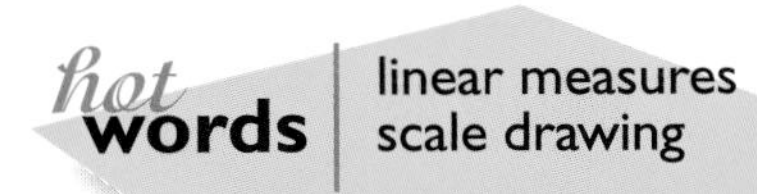

page 37

PHASE TWO

August 5, 1706

Having been condemned by nature and fortune to a restless life, I took a post on the Antelope, a trading ship.

I had been spending much time on deck, until a storm suddenly came upon us.

The storm grew steadily worse. The ship sank on a reef, and I became separated from my crew. Pushed toward shore by the tide, I fell exhausted upon the softest and shortest grass I had ever seen. But before I could examine it further, I fell into a deep sleep.

When I awoke, I found my arms and legs strongly fastened to the ground. Even my hair was tied down. My frustration gave way to surprise as I saw who had done this. Hundreds of them were scattered on the ground around me. They looked like men in every way but one—they were a mere six inches tall.

Lemuel Gulliver

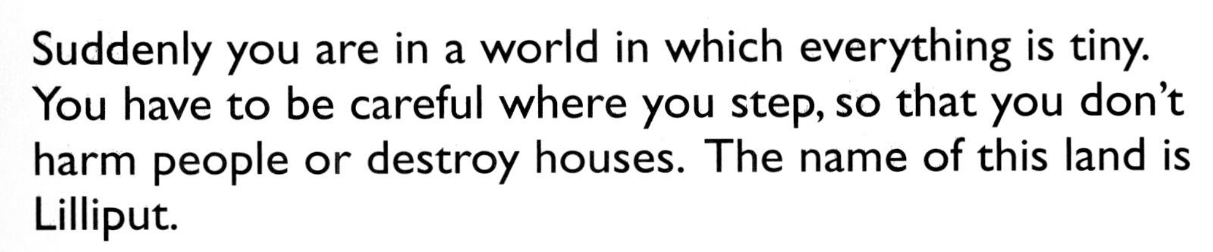

Suddenly you are in a world in which everything is tiny. You have to be careful where you step, so that you don't harm people or destroy houses. The name of this land is Lilliput.

In Phase Two, you will use a scale factor that makes things smaller. You will compare the different ways of measuring things using inches and feet, centimeters and meters and the measurement units used in Lilliput.

Lilliput

WHAT'S THE MATH?

Investigations in this section focus on:

DATA COLLECTION

- Gathering information from a story and pictures
- Organizing data to find patterns

MEASUREMENT and ESTIMATION

- Measuring accurately using fractions
- Comparing the U.S. customary and metric systems of measurement
- Estimating the sizes of objects
- Exploring area and volume measurements

SCALE and PROPORTION

- Working with a scale factor that reduces the sizes of objects
- Applying the scale factor to predict sizes of objects and to create a three-dimensional scale model
- Exploring the effect of rescaling on area and volume

5 Sizing Up the Lilliputians

DETERMINING A SCALE FACTOR LESS THAN ONE

Gulliver is swept overboard in a storm at sea and wakes up in a new land. He is the captive of tiny people in the land of Lilliput. How are the sizes of things in Lilliput related to sizes in Ourland? Clues in the journal will help you find out how small things are in Lilliput.

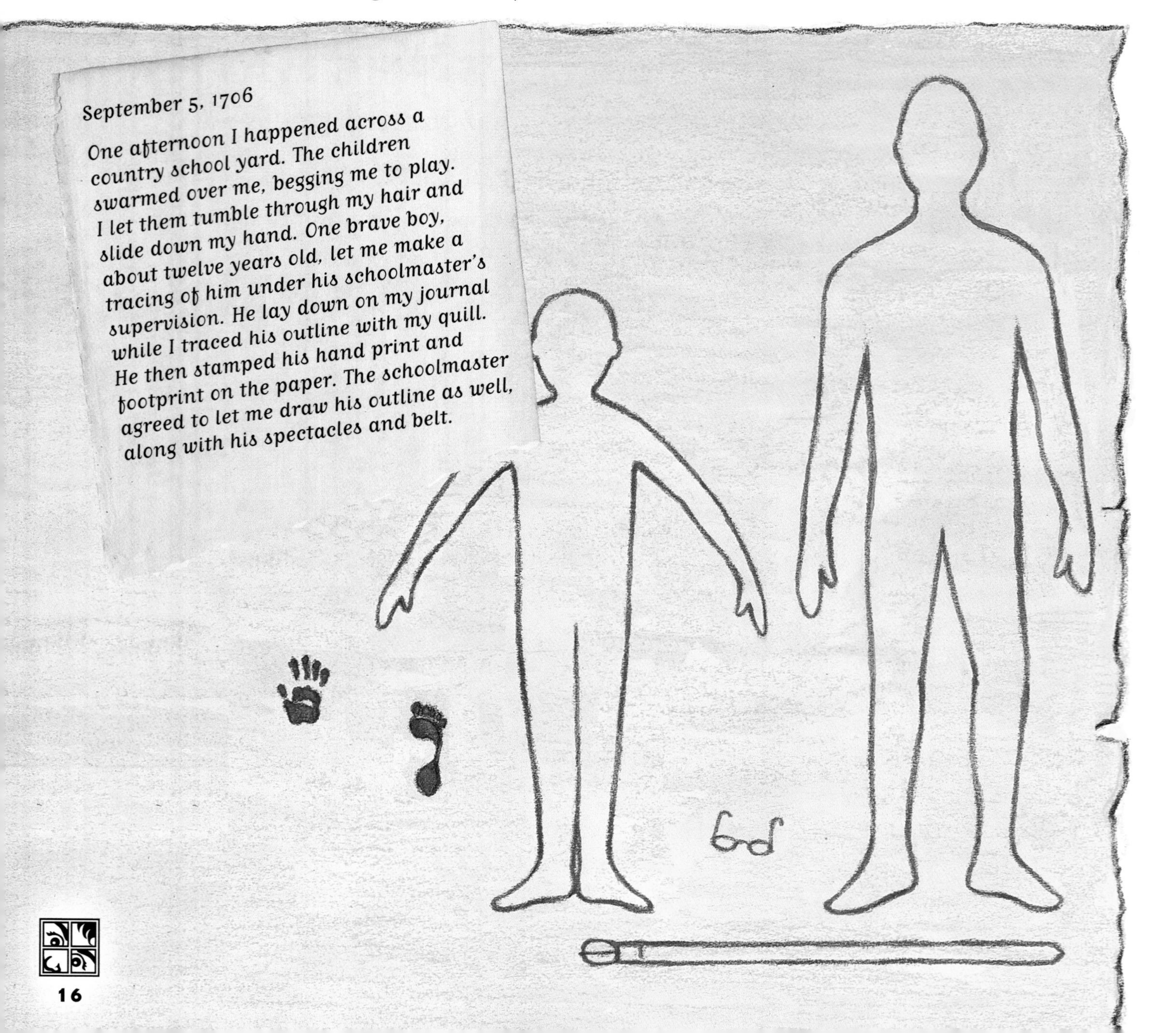

September 5, 1706

One afternoon I happened across a country school yard. The children swarmed over me, begging me to play. I let them tumble through my hair and slide down my hand. One brave boy, about twelve years old, let me make a tracing of him under his schoolmaster's supervision. He lay down on my journal while I traced his outline with my quill. He then stamped his hand print and footprint on the paper. The schoolmaster agreed to let me draw his outline as well, along with his spectacles and belt.

Create a Chart to Compare Sizes

How are sizes of things in Lilliput related to sizes in Ourland?

Make a scale chart with three columns. Column 1 is for the name of an object. Column 2 is for the size of the object in Lilliput. Column 3 is for the size in Ourland. Use a ruler to measure the tracings.

1. Use the words and tracings in the story to record the name of the object and its Lilliputian measurements on the chart. Measure or estimate the size of the same object in Ourland.
2. Use the information in your scale chart to find a scale factor that shows how sizes in Lilliput are related to sizes in Ourland.
3. Estimate or measure the sizes of some more objects in Ourland. Add these objects and their Ourland measurements to the chart. Find the size each object would be in Lilliput and add that information to the chart.

Do you think the Lilliputian student in the tracing is tall, short, or average-size in a Lilliputian sixth-grade class?

Write About Estimation Strategies

Write about what you did and learned as you investigated sizes of things in Lilliput.

- Describe the measurement and estimation strategies you used to find the sizes of things in Ourland and Lilliput. Show how you used the scale factor to complete your chart.
- What did you discover about finding the average size of an object?

6 Glum-gluffs and Mum-gluffs

MEASURING WITH NONSTANDARD UNITS

The same object can be measured in different units of measurement. Inches and feet are units in the U.S. customary system of measurement. Centimeters and meters are units in the metric system. How do these units compare to the units used in Lillilput?

October 5, 1706

During dinner, the King and Queen told stories about their country and people, and I told stories of mine. The King found it especially hard to believe that he, one of the tallest men in his land, would be no bigger than a child's doll in mine. He informed me that he was $8\frac{1}{2}$ glum-gluffs tall, and that the Queen was 6 glum-gluffs tall. When I inquired what a glum-gluff was, he replied that it was $\frac{1}{20}$ of a mum-gluff. He then kindly agreed to have his steward mark the length of 1 glum-gluff in my journal.

———— 1 glum-gluff

Measure an Object in Different Systems

Choose an object that you added to the Lilliput scale chart you made in Lesson 5.

1. Use the measurements recorded in the chart to make an accurate, life-size drawing of the object in Lilliput.
2. Use a metric ruler to measure the drawing in metric units (centimeters). Write the metric measurements on the drawing.
3. Calculate what the measurements of the drawing would be in the Lilliputian units of glum-gluffs. Write the Lilliputian measurements on the drawing.
4. Compare the object in the drawing to any object in Ourland that would be about the same size. Write the name of the Ourland object on the drawing.

How do the units used in different measurement systems compare?

Compare Measurement Systems

Write a letter to the King and Queen of Lilliput. Compare the measurement systems of Ourland and Lilliput. Make sure your letter answers the following questions:

- When would you prefer to use the U.S. customary system of measurement? When would you prefer to use the metric system?
- Would you ever prefer to use glum-gluffs and mum-gluffs? Why?
- Suppose the people in Lilliput were going to adopt one of our measurement systems. Which one would you recommend to them? Why or why not?

standard measurement
measurement units

page 39

7 Housing and Feeding Gulliver

SOLVING PROBLEMS INVOLVING AREA AND VOLUME

Gulliver's needs for food and shelter in Lilliput present some interesting problems. These problems involve area and volume. In the last phase, you solved problems in one dimension. Now you will extend your work with the Lilliputian scale factor to solve problems in two and three dimensions.

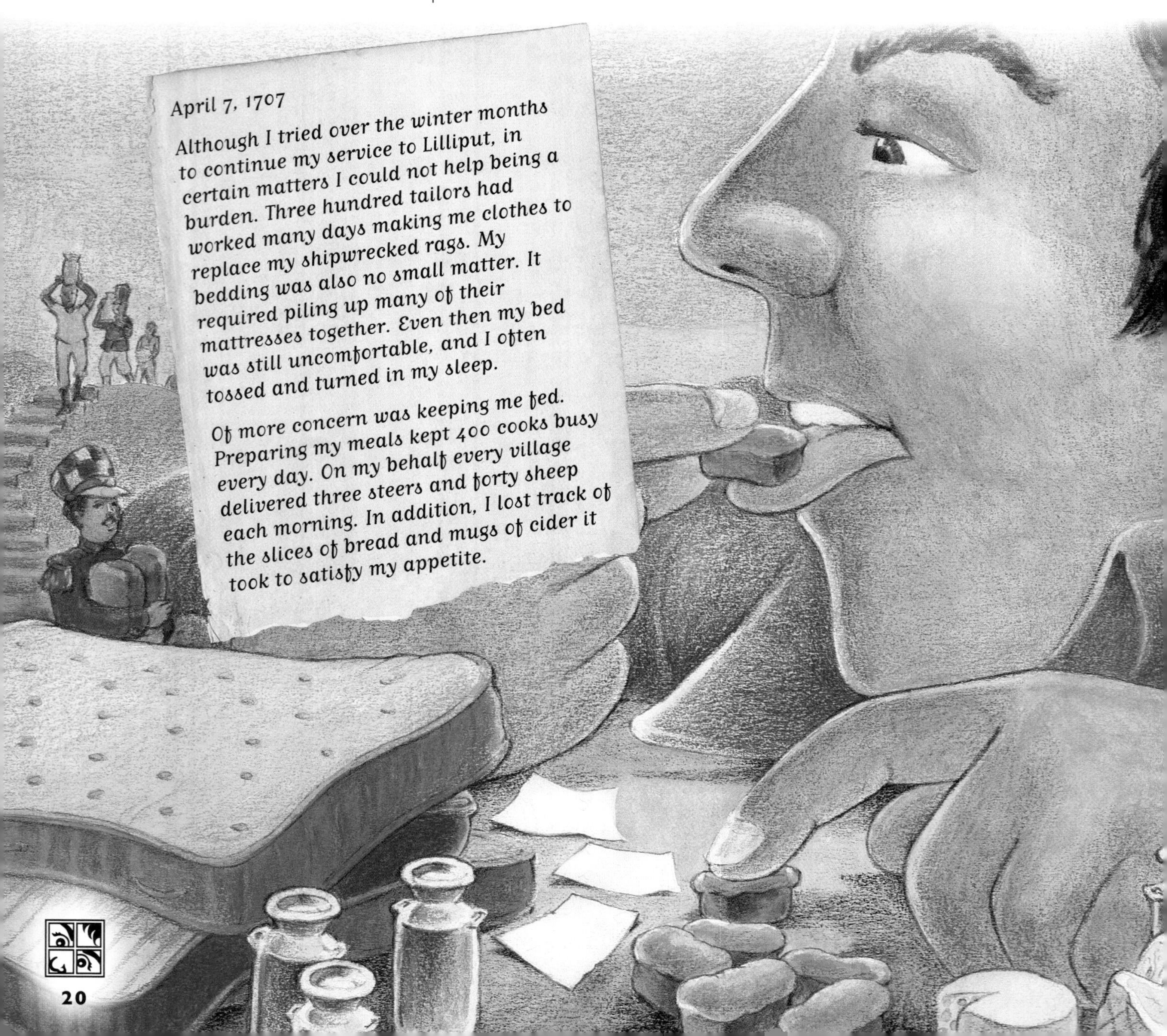

April 7, 1707

Although I tried over the winter months to continue my service to Lilliput, in certain matters I could not help being a burden. Three hundred tailors had worked many days making me clothes to replace my shipwrecked rags. My bedding was also no small matter. It required piling up many of their mattresses together. Even then my bed was still uncomfortable, and I often tossed and turned in my sleep.

Of more concern was keeping me fed. Preparing my meals kept 400 cooks busy every day. On my behalf every village delivered three steers and forty sheep each morning. In addition, I lost track of the slices of bread and mugs of cider it took to satisfy my appetite.

Estimate to Solve Area and Volume Problems

Estimate how many of the Lilliputian objects Gulliver needs. Then make a Lilliputian-size model of one of the four objects.

1. How many Lilliputian-size mattresses would Gulliver need to make a bed? How should Gulliver arrange those mattresses to make a comfortable-size bed?
2. How many sheets of Lilliputian paper would need to be taped together to made one sheet of writing paper for Gulliver?
3. At home, Gulliver would eat two loaves of bread each week. How many Lilliputian loaves of bread would Gulliver need each week?
4. At home, Gulliver drank 3 cups of milk each day. How many Lilliputian-size quarts of milk would Gulliver need each day?

How can you use estimation to solve problems in two and three dimensions?

Use a Model to Check an Estimate

Describe in writing how you can use your Lilliputian-size model to check your estimate. Include a sketch with measurements to show your thinking.

8 Seeing Through Lilliputian Eyes

WRITING ABOUT AREA AND VOLUME

Imagine yourself in Lilliput. What objects would you bring with you? How would the Lilliputians describe these objects? You will use what you have learned about scale in one, two, and three dimensions when you write a story describing your own adventures in Lilliput.

Gulliver's Pocket Contents:

1. One great piece of coarse cloth, large enough to be a carpet for your Majesty's chief Room of State
2. A great bundle of white thin substances, folded one over another, about the thickness of three men, tied with a strong cable and marked with black figures, with every letter almost half as large as the palm of our hands
3. A long pole from the back of which extended 20 shorter poles, resembling the palace railings
4. Several round flat pieces of yellow and silver metal, of different bulk, some so large and heavy that my comrade and I could hardly lift them
5. Some wonderful kind of globe-like engine, part silver and part transparent metal, with a loud noise like the sound of a water-mill, attached by a great silver chain

Write a Story Using 3-D Measurements

Write a believable story using three-dimensional measurements. You will need to figure out the correct length, width, and height of the objects you describe.

1. Imagine a place in Lilliput. Describe at least one adventure that could happen to you there.
2. Describe the measurements of at least three objects found in the place. You could include an Ourland object in your story for comparison.
3. Include a conversation with a Lilliputian that compares the sizes of the objects in the story to the same objects in Ourland.
4. Record and check all of your measurements.

How can you describe a three-dimensional Lilliputian world?

Describe Rescaling Strategies

Summarize how you determined the length, width, and height of the three objects described in your story.

- Make a table, list, or drawing showing the length, width, and height of each object in both Ourland and Lilliput.
- Explain the methods you used to estimate or measure each object. Show how you rescaled it using the scale factor.

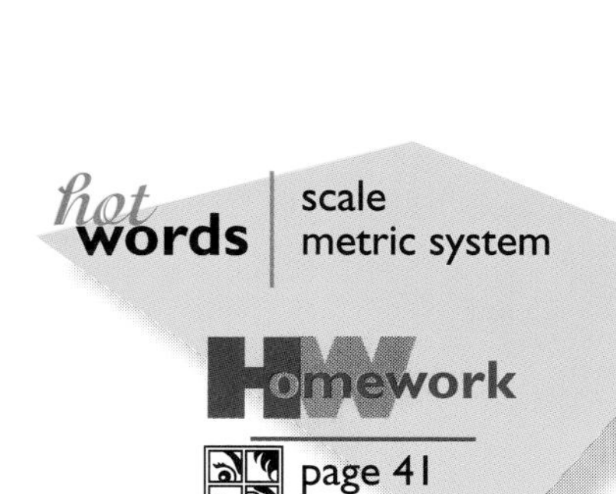

PHASE THREE

September 13, 1708

It has now been some months since my journal was published. News of my travels has prompted a greater response than I expected. I must confess, too, a certain pleasure at the attentions I have received from my publisher. He has pronounced himself well-pleased with the sales figures, and credits both my nature and my writing for this success.

Although I did not anticipate it, I have recently received much correspondence from my readers. Pictures and letters have come from nearby countries—France, Spain and Holland—as well as more distant lands. It is my dream that someday all my treasures and adventures may be shared with the public as part of a glorious exhibit, but, alas, I fear it will not happen in my lifetime.

Your friend,

Lemuel Gulliver

Imagine you are in charge of a special exhibit about *Gulliver's Worlds*. How would you show the sizes of the different lands he visited? In this final phase, you will explore size relationships in different lands, both big and small. You will find ways to show how the sizes compare in length, area, and volume. Finally, your class will create displays of life-size objects from one of *Gulliver's Worlds*.

Lands of the Large and Lands of the Little

WHAT'S THE MATH?

Investigations in this section focus on:

DATA COLLECTION

- Gathering information from pictures
- Creating displays to show size relationships

MEASUREMENT and ESTIMATION

- Measuring accurately using fractions
- Exploring area and volume measurements

SCALE and PROPORTION

- Finding scale factors that describe relationships among sizes
- Enlarging and reducing the sizes of objects according to scale factors
- Creating 2-D scale drawings
- Creating a 3-D scale model
- Exploring the effects of rescaling on area and volume

Lands of the Large

REPRESENTING SIZE RELATIONSHIPS

Ourland Museum needs a display that compares the sizes of objects from Lands of the Large to Ourland. Can you figure out the scale factor for each of the Lands of the Large? Can you find a way to show the size relationships among the different lands?

How large is a life-size face in each of the Lands of the Large?

Investigate Proportions of Faces

Compare the objects in the photos to find the scale factor. The smaller object is always from Ourland. When you are finished, check to make sure your scale factor is correct before you do the following group investigation.

1. As a group, select one of the Lands of the Large for this investigation. Have each member of your group draw a different feature of a face from your group's land.
2. As a group, arrange the features to form a realistic face. Check to see if your measurements are correct and the features are in proportion. Work together to draw the outline of the face.

How tall would a person in your group's Land of the Large be?

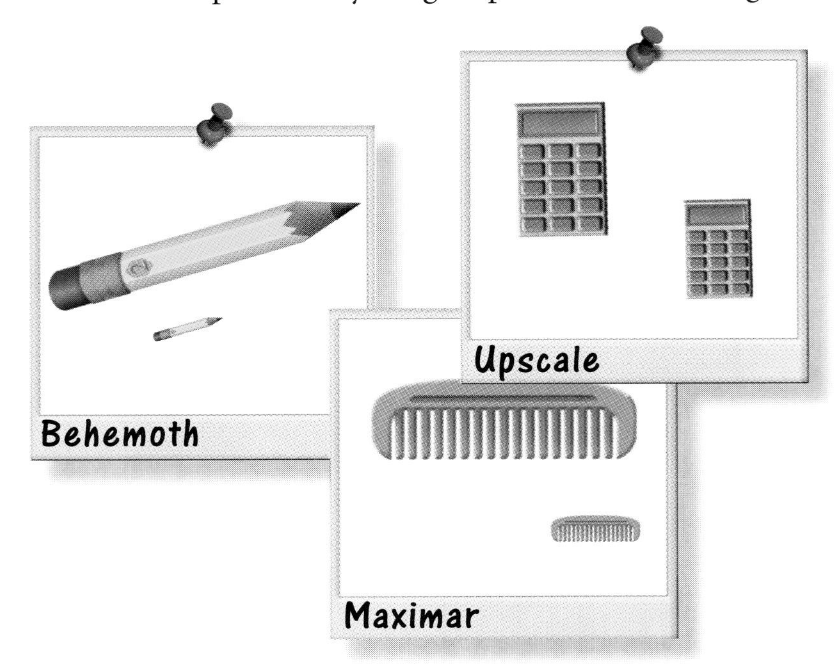

Represent Size Relationships

How do things in the Lands of the Large compare in size to things in Ourland?

Use the scale drawing of an Ourland face on this page to make a simple scale drawing of a face from each of the Lands of the Large. Organize your drawings into a visual display to show size relationships.

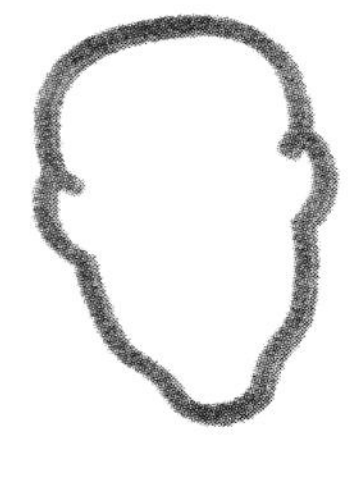

1. Measure the scale drawing of the Ourland face.
2. Use the scale factors from the Lands of the Large to make a scale drawing of a face from each land. You do not need to draw in the features.
3. Write the name of the land and the scale factor compared to Ourland next to each drawing.
4. Organize your drawings into a display of size relationships that compares the sizes of faces from different lands and shows how they are related.

Describe a Scale Factor for Brobdingnag

Compare the sizes of things in each of the Lands of the Large to the sizes of things in Brobdingnag. Use this to explain how the scale factor describes size relationships.

- Figure out the scale factor for each land compared to Brobdingnag. Write it next to the scale drawing from that land.
- Describe in writing how you figured out the scale factor. Tell why it is different from the Ourland scale factor.

10 Lands of the Little

PREDICTING SCALE SIZE

Can you find the mistakes in the Lands of the Little display? Here you will correct the scale drawings and create a chart that you can use to find the size of any object in a Land of the Little.

How do things in the Lands of the Little compare in size to things in Ourland?

Compare Objects in the Lands of the Little

Measure each pair of scale drawings on this page. The larger object is always from Ourland. Does the size relationship for each Land of the Little object match the scale factor below it?

1. As a group, choose an Ourland object from your classroom. Draw your object on a piece of paper.
2. Use each of the four scale factors below to draw a new picture of your object. Now, which scale drawings below do you think are incorrect?

Quarterville 1:4 (.25:1)

Micropolus 3:8 (.375:1)

Small Town 2:3 (.67:1)

Dimutia 1:6 (.167:1)

Create a Table to Show Size Relationships

Make a table that shows how objects in the Lands of the Little compare in size to objects in Ourland. Record the size an object would be in other lands if you knew its size in Ourland.

1. Fill in the names of the Lands of the Little at the top of each column. Mark the Ourland measurements (100 inches, 75 inches, 50 inches, 25 inches, 10 inches) in the Ourland column.
2. Figure out how big an object would be in the Lands of the Little for each of the Ourland measurements. Mark the Lands of the Little measurements in the appropriate columns.
3. Find a way to use your group's scale drawings to check that your table is correct.

How could you show the size relationships of things in different lands?

Ourland	Lilliput	Dimutia	Quarterville	Micropolus	Small Town
100 inches					
75 inches					
50 inches					
25 inches					
10 inches					

Write a Guide for Using the Table

Explain how you can use your table to answer each question.

- If an object is 80 inches long in Ourland, how long would it be in each of the other lands?
- If an object is 5 inches long in Lilliput, how long would it be in Ourland?
- If an object is 25 inches long in Small Town, how long would it be in the other lands?

11 Gulliver's Worlds Cubed

RESCALING IN ONE, TWO, AND THREE DIMENSIONS

The *Gulliver's Worlds* group exhibit needs a finishing touch. The exhibit needs to show how rescaling affects area and volume. How do length, area, and volume change when the scale of something changes? Can you create a display that will help visitors understand this?

How does a change in scale affect measurements of length, area, and volume?

Investigate Cube Sizes in Different Lands

Use the information on this page to figure out a way to use Ourland cubes to build a large cube at each of the following scale factors: 2 : 1, 3 : 1, 4 : 1.

1. Record how many Ourland cubes make up each large cube.
2. Estimate how many cubes it would take to make a Brobdingnag cube (scale factor = 12 : 1)

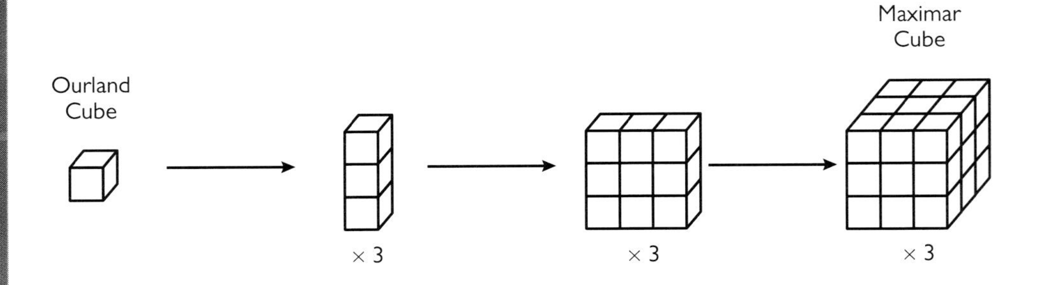

Comment Card

If Maximar is 3 times bigger than Ourland, why does it take more than 3 cubes from Ourland to make a cube in Maximar?

Collect Data for Two and Three Dimensions

Use your cubes to collect size information.

1 Organize the information into a table that answers the following questions:

a. What is the scale factor of the cube?

b. How many Ourland cubes high is one edge of this cube?

c. How many Ourland cubes are needed to cover one face of this cube completely?

d. How many Ourland cubes are needed to fill this cube completely?

Scale Factor	How Many Cubes Long Is an Edge? (length)	How Many Cubes Cover a Face? (area)	How Many Cubes Fill the Cube? (volume)
2:1			
3:1			
4:1			
5:1			
10:1			
25:1			

2 Find a rule that can predict how big a cube would be for each of the following scales. Then add the information to your table:

2.5:1 6:1 20:1 100:1

How can you predict how much each measurement will change when you rescale?

Write About Scale, Area, and Volume

Write down the set of rules you used to complete your table. Make sure that someone else could apply your rules to any scale factor.

1 Explain how your rules work.

2 Make a diagram showing how to use the rules to predict the following:

a. The length of one edge of a cube

b. The area of one face of a cube

c. The volume of a cube

exponents
cubic centimeter

page 44

12 Stepping into Gulliver's Worlds

FINAL PROJECT

A life-size display in the correct scale and proportion can make you feel like you have stepped into another world. You will help the Ourland Museum create a life-size display of one of the lands in *Gulliver's Worlds.* The goal is for museum visitors to get involved with your display.

What would it look like if you stepped into one of the lands in *Gulliver's Worlds*?

Create a Display Using Accurate Dimensions

Choose one of the lands in *Gulliver's Worlds.* Create a display and tour that compares the sizes of things in that land to those in Ourland.

1. Create at least three objects in one, two, or three dimensions to use in the display.
2. Write a short tour that describes the measurements of the objects in the display and compares them to sizes in Ourland. Describe and label the areas and volumes of the objects.
3. Find a way for visitors to get involved with the display.
4. Include your writing from previous lessons and charts to help museum visitors understand the scale of the land.

Gulliver Show Opens At The Ourland Living Museum

by Jonathan Swift
Ourland News Correspondent

The *Gulliver's Worlds* exhibit at the Ourland Living Museum is an exciting journey to new lands. From my entrance, where I was met by a huge smile from a life-size Brobdingnag face, to the carefully crafted scale drawings of the Lands of the Little gallery, the exhibit showed this reporter what it would be like to actually live in the worlds that Gulliver explored hundreds of years ago in his famous journal.

Review a Display

You will be reviewing a classmate's display and presentation. As you review the exhibit, write down the scale factor and as many measurements as you can. Use the following questions to help write your review:

1. What parts of the display look life-size?
2. How did you check that the sizes of the objects in the display were correct?
3. How does the presentation describe linear, area, and volume measurements?
4. How does the presentation compare sizes to those in Ourland?
5. Would you add or change anything to make the display more believable?

How would you evaluate your own display?

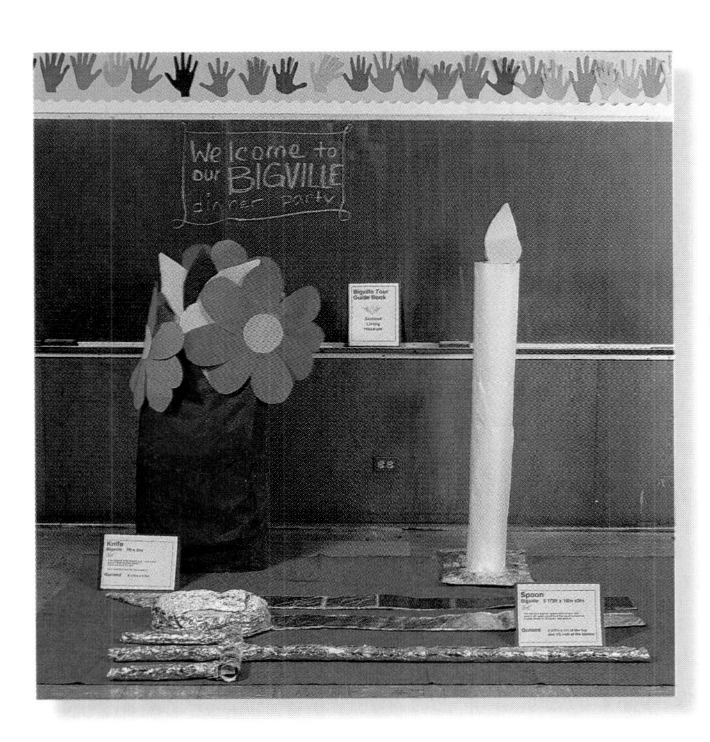

The Sizes of Things in Brobdingnag

Applying Skills

Fill in the missing height conversions to complete the chart.

	Name	Height (in.)	Height (ft and in.)	Height (ft)
	Marla	49″	4′1″	$4\frac{1}{12}$′
1.	Scott	56″		
2.	Jessica		4′7″	
3.	Shoshana	63″		
4.	Jamal	54″		
5.	Louise		4′11″	
6.	Kelvin	58″		
7.	Keisha		5′2″	
8.	Jeffrey		4′2″	

9. List the names in height order from tallest to shortest.

10. The scale factor of Giantland to Ourland is 11:1. That means that objects in Giantland are 11 times the size of the same objects in Ourland. Figure out how large the following Ourland objects would be in Giantland:

a. a tree that is 9 ft tall

b. a man that is 6 ft tall

c. a photo that is 7 in. wide and 5 in. high

11. The scale factor of Big City to Ourland is 5:1. That means that objects in Big City are 5 times the size of the same objects in Ourland. How large would each of the Ourland objects from item **10** be in Big City?

Extending Concepts

12. Duane made an amazing run at the football game Friday night.

Examine the diagram below and give the distance of the play in:

a. yards **b.** feet **c.** inches

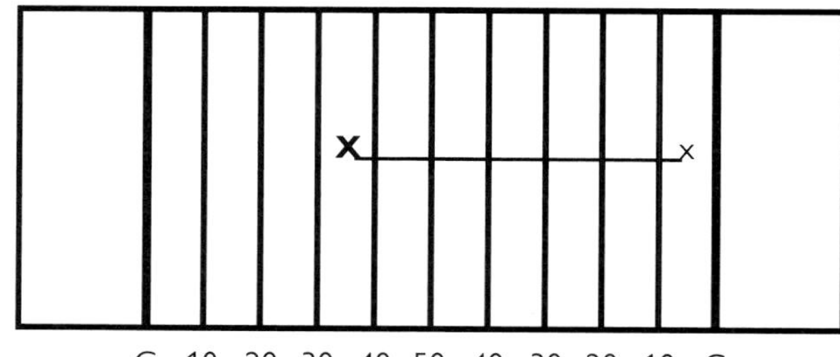

HINT: From goal line to goal line in a football field is 100 yards.

Making Connections

13. Answer this Dr. Math letter:

Dear Dr. Math,
Today in science class we were using microscopes. The lenses were 10×, 50×, and 100×. I think there is a way scale factor applies to what I see and what the actual size is. Is that true? If so, could you please explain?
William Neye

A Life-Size Object in Brobdingnag

Applying Skills

Reduce these fractions to lowest terms.

1. $\frac{21}{49}$ **2.** $\frac{33}{126}$ **3.** $\frac{54}{81}$

4. $\frac{28}{48}$ **5.** $\frac{15}{75}$ **6.** $\frac{10}{18}$

7. $\frac{126}{252}$ **8.** $\frac{8}{24}$ **9.** $\frac{16}{12}$

10. $\frac{64}{6}$

Follow the instructions to describe each relationship in a different way.

11. Write $\frac{10}{1}$ as a ratio.

12. Write 4:1 as a fraction.

13. Write "2 to 1" as a fraction.

14. Write out 6:1 in words.

15. Write $\frac{8}{1}$ as a ratio.

Extending Concepts

16. The height of a blade of grass in a giant-size display is $4\frac{2}{3}$ ft. The blade of grass in your yard is 4 in. high. What is the scale factor?

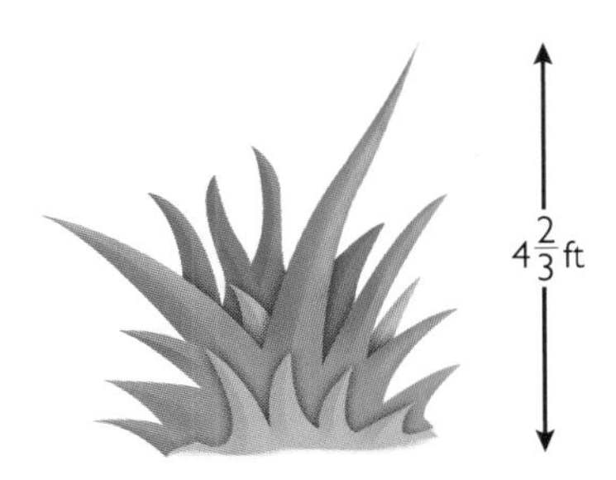

17. The scale factor of Vastland to Ourland is 20:1. That means that objects in Vastland are 20 times the size of the same objects in Ourland. Figure out how large the following Ourland objects would be in Vastland:

a. a car that is $4\frac{1}{2}$ ft high and 8 ft long

b. a building that is 23 yd high and 40 ft long

c. a piece of paper that is $8\frac{1}{2}$ in. wide and 11 in. long

Making Connections

18. The science class is creating insects that are larger than life. First they will study the ant. The queen ant that they have to observe is $\frac{1}{2}$ in. long. The large model they create will be 5 ft long. Mr. Estes wants to have a ladybug model created too. Tasha found a ladybug and measured it at $\frac{1}{8}$ in. long.

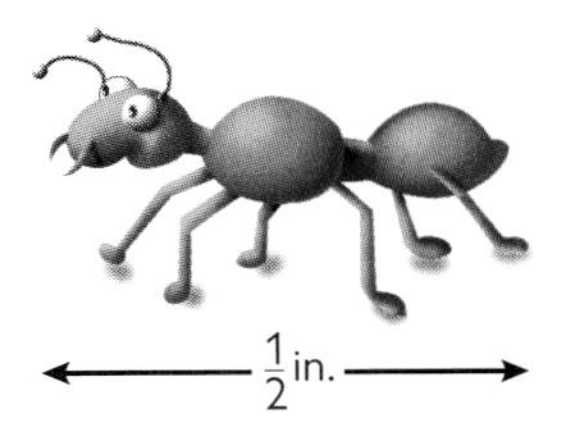

a. What is the scale factor for the ant model?

b. How big will the ladybug model be if the same scale factor is used?

How Big Is "Little" Glumdalclitch?

Applying Skills

Reduce these fractions to the lowest terms.

1. $\frac{25}{75}$ **2.** $\frac{69}{23}$ **3.** $\frac{1,176}{21}$

4. $\frac{16}{4}$ **5.** $\frac{36}{30}$ **6.** $\frac{49}{14}$

7. $\frac{24}{3}$ **8.** $\frac{54}{18}$ **9.** $\frac{8}{12}$

Convert these fractions to like measurement units and then reduce each fraction to show a size relationship. See if you can make each fraction into a scale factor.

10. $\frac{3 \text{ in.}}{4 \text{ ft}}$ **11.** $\frac{8 \text{ in.}}{2 \text{ yd}}$ **12.** $\frac{440 \text{ yd}}{\frac{1}{2} \text{ mi}}$

13. $\frac{18 \text{ in.}}{1 \text{ yd}}$ **14.** $\frac{2 \text{ yd}}{12 \text{ ft}}$

Extending Concepts

15. Ali is writing a script for a new movie in which aliens that are 3 times the size of humans (3:1) take the game of football back to their home planet. HINT: U.S. regulation football fields measure 100 yards from goal line to goal line.

a. How long is the aliens' field in yards?

b. How long is the aliens' field in inches?

c. How long is the aliens' field in feet?

16. The scale factor of Jumbolia to Ourland is 17:1. That means that objects in Jumbolia are 17 times the size of the same objects in Ourland. Figure out how large the following Ourland objects would be in Jumbolia:

a. a radio that is $4\frac{1}{2}$ in. high, 8 in. long, and $3\frac{1}{2}$ in. wide

b. a rug that is $6\frac{1}{2}$ ft wide and $9\frac{3}{4}$ ft long

c. a desk that is $2\frac{1}{3}$ ft high, 3 ft long, and $2\frac{1}{2}$ ft wide

Making Connections

17. Provide the scale factor for the following map by measuring the distance with a ruler. The distance from the library to the school is $2\frac{1}{2}$ miles.

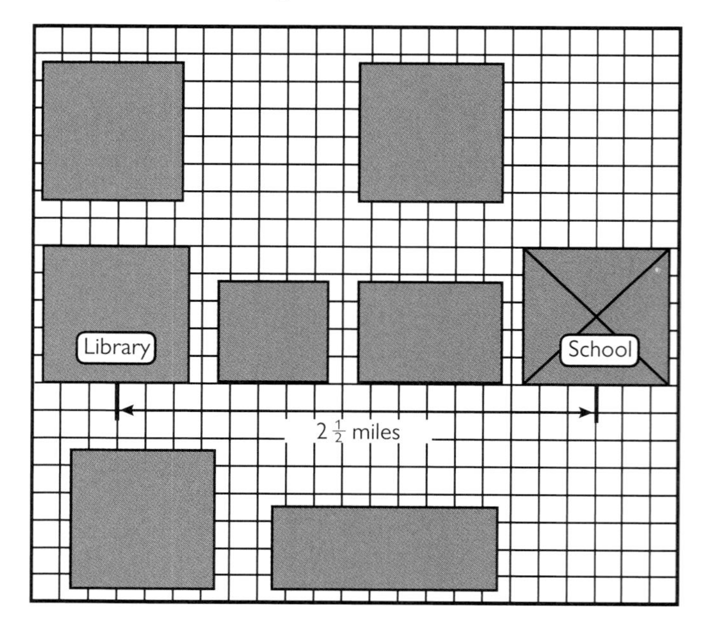

Telling Tales in Brobdingnag

Applying Skills

Convert these ratios to like measurement units and then reduce the fraction to create a scale factor.

Example $\frac{1}{2}$ yd: 6 in. $= \frac{\frac{1}{2}\text{ yd}}{6\text{ in.}} = \frac{18\text{ in.}}{6\text{ in.}} = \frac{3\text{ in.}}{1\text{ in.}}$ = scale factor 3:1

1. $\frac{3}{4}$ ft:3 in.

2. 6 ft: $\frac{1}{3}$ yd

3. $\frac{1}{2}$ mi : 528 ft

4. $2\frac{1}{2}$ yd: $\frac{1}{4}$ ft

5. 14 in : $\frac{7}{12}$ ft

6. $\frac{1}{6}$ yd:2 in.

7. $\frac{1}{15}$ mi:16 ft

8. $4\frac{1}{12}$ ft:7 in.

9. 4 mi:1,760 yd

10. 4,392 in:6 ft

Extending Concepts

11. The scale factor of Mammothville to Ourland is 1 yd:1 in. That means if an object in Mammothville is one yard long, then the same object in Ourland would be only one inch long. Figure out how large the following Ourland objects would be in Mammothville:

a. a soda can that is 5 in. tall and $2\frac{1}{2}$ in. wide

b. a football field that is 100 yd long

c. a table that is $3\frac{1}{2}$ ft high, 4 ft wide, and 2 yd long

12. The scale factor of Colossus to Ourland is $\frac{1}{4}$ yd:3 in. That means if an object in Colossus is $\frac{1}{4}$ of a yard long, then the same object in Ourland would be only 3 inches long. How large would each of the Ourland objects from item **11** be in Colossus?

13. Match the following scale factors to the correct measurement units:

a. 3:1 **i.** 1 ft:1 in.

b. 5,280:1 **ii.** 1 mi:1 yd

c. 12:1 **iii.** 1 yd:1 ft

d. 1,760:1 **iv.** 1 mi:1 ft

Making Connections

14. In Humungoville the scale factor to Ourland is 4:1. Use the postage stamp from Ourland pictured below to draw a postage stamp for Humungoville. Make sure the length and width are at a scale factor of 4:1. You can be creative with the picture inside.

Sizing Up the Lilliputians

Applying Skills

Write each of the following decimals as a fraction.

Example $0.302 = \frac{302}{1,000}$

1. 0.2 **2.** 0.435

3. 0.1056 **4.** 0.78

5. 0.44 **6.** 0.025

7. 0.9 **8.** 0.5002

9. 0.001 **10.** 0.67

Write each decimal in words.

Example 0.5 = five tenths

11. 0.007 **12.** 0.25

13. 0.3892 **14.** 0.6

15. 0.04

16. The scale factor of Pint-Size Place to Ourland is 1:11. That means that objects in Ourland are 11 times the size of the same objects in Pint-Size Place. Figure out about how large the following Ourland objects would be in Pint-Size Place:

a. a house that is 15 ft high, 33 ft wide, and 60 ft long

b. a train that is 363 ft long and 20 ft high

c. a woman who is 5 ft 6 in. tall

Extending Concepts

17. Measure the height of each picture. Compare the sizes of the pictures and determine the scale factor. What is the scale factor when:

a. the larger picture is 1?

b. the smaller picture is 1?

Making Connections

18. The regulation size of a soccer field varies from the largest size, 119 m × 91 m, to the smallest size allowed, 91 m × 46 m. What is the difference in the perimeters of the two field sizes? How do you think the difference in perimeters affects the game?

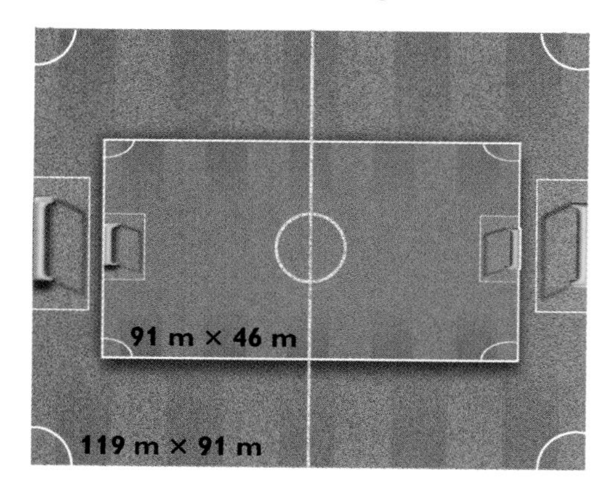

Glum-gluffs and Mum-gluffs

Applying Skills

Complete the following table showing equivalencies in the metric system.

	mm	cm	dm	m	km
1.				1,000	1
2.	1,000	100	10	1	
3.			1		
4.		1			
5.	1				

Supply the missing equivalent.

6. 42 dm = _____ m

7. 5 cm = _____ m

8. 0.5 m = _____ cm

9. 0.25 cm = _____ mm

10. 0.45 km = _____ m

11. 1.27 m = _____ dm

12. 24.5 dm = _____ cm

13. 38.69 cm = _____ m

14. 0.2 mm = _____ cm

15. 369,782 mm = _____ m

16. 0.128 cm = _____ mm

17. 7.3 m = _____ dm

Extending Concepts

18. Place the following measurements in height order from shortest to tallest.

- 1967 mm
- 0.0073 km
- 43.5 cm
- 0.5 m
- 7 dm

Making Connections

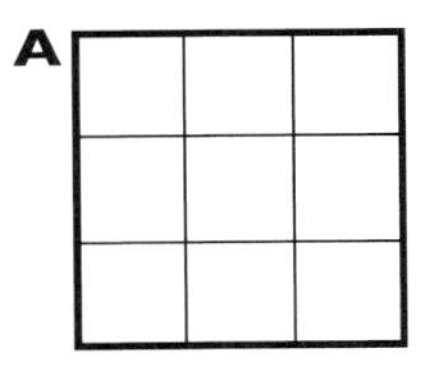

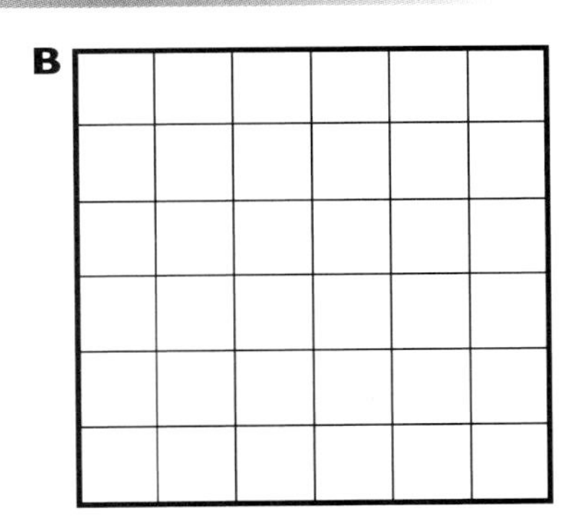

19. Count the smaller squares to figure out the sizes of squares A and B in square units.

a. What are the perimeter and the area of square A?

b. What are the perimeter and the area of square B?

20. Compare the two perimeters and the two areas. Describe each size relationship using a scale factor.

Housing and Feeding Gulliver

Applying Skills

Convert these fractions to like measurement units.

Example $\frac{4 \text{ m}}{4 \text{ cm}} = \frac{400 \text{ cm}}{4 \text{ cm}}$

1. $\frac{43 \text{ cm}}{43 \text{ mm}}$ **2.** $\frac{5 \text{ m}}{5 \text{ cm}}$ **3.** $\frac{6 \text{ km}}{6 \text{ m}}$

Use your answers from items 1–3 to show a scale factor that is less than one. HINT: Reduce the larger number to one.

Example $\frac{400 \text{ cm} \div 400}{4 \text{ cm} \div 400} = \frac{1}{0.01} = 1{:}0.01$

4. $\frac{43 \text{ cm}}{43 \text{ mm}}$ **5.** $\frac{5 \text{ m}}{5 \text{ cm}}$ **6.** $\frac{6 \text{ km}}{6 \text{ m}}$

7. The scale factor of Teeny Town to Ourland is 1:6. That means that objects in Ourland are 6 times the size of the same objects in Teeny Town. Figure out how large the following Ourland objects would be in Teeny Town:

a. a book 30 cm high and 24 cm wide

b. a girl 156 cm tall

c. a table 1 m high, 150 cm wide, and 2 m long

Extending Concepts

8. Albert is using a scale factor of 3:1 for his school project. The height of the walls he measured are 3 m and the walls in the model he made are 1 m high. A 3-ft-high chair became a 1-ft-high chair in his project. Can he use both metric and U.S. customary measurement units in the same project? Why or why not?

9. The scale factor of Itty-Bittyville to Ourland is 1:4. That means that objects in Ourland are 4 times the size of the same objects in Itty-Bittyville. Estimate the sizes of each of the following Itty-Bittyville objects and find an object in Ourland that is about the same size:

a. an Itty-Bittyville textbook

b. an Itty-Bittyville double bed

c. an Itty-Bittyville two-story building

d. an Itty-Bittyville car

10. Estimate the scale factor of Peeweeopolis to Ourland if the area of an Ourland postage stamp is equal to the area of a Peeweeopolis sheet of paper.

Making Connections

For items 11–13 use the figure below.

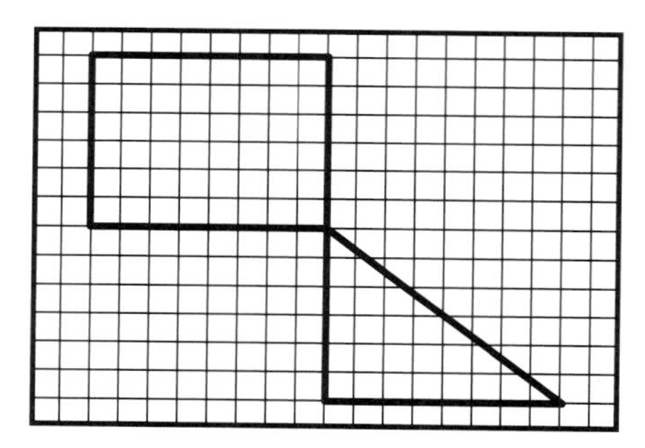

11. What is the area in square units of:

a. the rectangle? **b.** the triangle?

12. Enlarge each shape using a scale factor of 3:1. What is the area in square units of:

a. the rectangle? **b.** the triangle?

13. How did you figure out the area of each shape for items **11** and **12**?

Seeing Through Lilliputian Eyes

Applying Skills

Reduce the following fractions to the lowest terms.

Example $\frac{36}{42} = \frac{6}{7}$

1. $\frac{81}{63}$ **2.** $\frac{4}{24}$ **3.** $\frac{16}{20}$

4. $\frac{5}{50}$ **5.** $\frac{27}{36}$ **6.** $\frac{36}{48}$

7. $\frac{90}{120}$ **8.** $\frac{12}{10}$ **9.** $\frac{75}{100}$

10. $\frac{11}{33}$ **11.** $\frac{14}{21}$ **12.** $\frac{80}{25}$

13. $\frac{9}{18}$ **14.** $\frac{4}{12}$

Making Connections

17. In science-fiction movies, miniatures and scale-factor models are used to create many of the special effects. In one case, the special effects team created several different scale models of the hero's spaceship. The life-size ship that was built to use for the filming was 60 ft long. One scale model was 122 cm long by 173 cm wide by 61 cm high.

a. What was the scale factor?

b. What were the width and height of the life-size ship?

Extending Concepts

15. Measure the length and width of each of the following shapes. Which measurement system, metric or U.S. customary, would be the easiest to use to enlarge each object using a scale factor of 2 : 1? Why?

a.

b.

16. The scale factor of Miniopolis to Ourland is 1 : 7. That means that objects in Miniopolis are $\frac{1}{7}$ the size of the same objects in Ourland. Figure out how large the following Ourland objects would be in Miniopolis:

a. a building that is 147 ft high, 77 ft wide, and 84 ft long

b. a road that is 2 miles long

c. a piece of paper that is $8\frac{1}{2}$ in. by 11 in.

Lands of the Large

Applying Skills

In the following exercises provide equivalent decimals.

Example $\frac{1}{2} = 0.5$

1. $\frac{1}{20}$ **2.** $\frac{1}{3}$ **3.** $\frac{1}{4}$

4. $\frac{1}{5}$ **5.** $\frac{1}{10}$ **6.** $\frac{1}{8}$

7. $\frac{3}{4}$ **8.** $\frac{1}{7}$

9. The scale factor of Big City to Ourland is 6.5:1. That means that objects in Big City are 6.5 times the size of the same objects in Ourland. Figure out how large the following Ourland objects would be in Big City:

a. a tree that is 9 ft tall

b. a man that is 6 ft tall

c. a photo that is 7 in. wide and 5 in. long

Extending Concepts

10. Complete the following table by figuring out equivalent scale factors for each row.

	Decimals	Fractions	Whole Numbers
	1.5 : 1	$1\frac{1}{2}$: 1	3 : 2
a.	6.5 : 1		
b.		$8\frac{1}{4}$: 1	
c.			5 : 3

11. The scale factor of Big City to Hugeville is 3:2. That means that objects in Big City are 1.5, or $1\frac{1}{2}$, times the size of the same objects in Hugeville. How large would each of the Big City objects from item **9** be in Hugeville?

Making Connections

12. The scale factor is 5:1 for a giant ice cube in comparison to the school cafeteria's ice cubes.

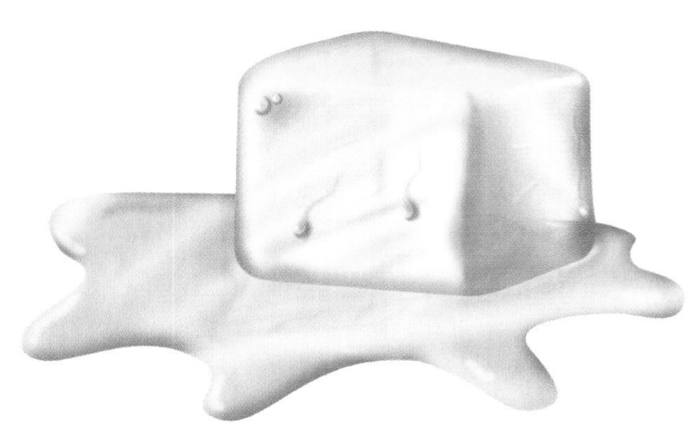

a. Draw a picture that shows how many cafeteria ice cubes you would have to stack high, wide, and deep to build a giant ice cube.

b. What is the total number of cafeteria ice cubes it would take to fill one giant ice cube?

Lands of the Little

Applying Skills

Complete the following chart by supplying the missing equivalents as decimals or fractions.

	Fractions	Decimals
1.	$\frac{1}{2}$	
2.		0.25
3.	$\frac{2}{3}$	
4.		0.7
5.	$\frac{3}{4}$	
6.		0.05
7.	$\frac{3}{8}$	
8.		0.125

Reduce the scale factor to a fraction. HINT: Divide each number by the largest number.

Example $10:7 = \frac{10}{10}:\frac{7}{10} = 1:\frac{7}{10}$

9. 3:2 **10.** 4:3 **11.** 5:3

Extending Concepts

12. The scale factor for Giantland to Ourland is 10:1. What is the scale factor from Ourland to Giantland? Write the scale factor using a decimal or fraction for Ourland to Giantland.

13. The scale factor of Wee World to Ourland is 0.5:1. That means that objects in Wee World are 0.5 the size of the same objects in Ourland. What is another way to write this scale factor without using a decimal?

14. Using the scale factor from item **13,** figure out how large the following Ourland objects would be in Wee World:

a. a house that is 15 ft high, 35 ft wide, and 60 ft long

b. a train that is 360 ft long and 20 ft high

c. a woman who is 5 ft 4 in. tall

Making Connections

15. Answer this Dr. Math letter:

Dr. Math,

When I was doing problems 9–11 in today's homework, my friend said there was a pattern between the whole-number scale factor and the fraction scale factor. I don't see it. Can you please explain it to me? Could I use this pattern to rescale objects more efficiently?

D.S. Mall

Gulliver's Worlds Cubed

Applying Skills

Complete the following chart with equivalent expressions.

	Exponent	Arithmetic Expression	Value
	3^3	$3 \times 3 \times 3$	27
1.	2^2		
2.		$4 \times 4 \times 4$	
3.			25
4.	6^2		
5.			49
6.	8^3		
7.			81
8.	10^2		
9.	5^3		
10.		$6 \times 6 \times 6$	

Tell whether each unit of measurement would be used for area or volume.

11. yd^2 (square yard)

12. cm^3 (cubic centimeter)

13. m^2 (square meter)

14. in^2 (square inch)

15. ft^3 (cubic feet)

16. mm^3 (cubic millimeter)

Extending Concepts

17. Concrete To Go is going to pour a patio 4 yd long, 4 yd wide, and $\frac{1}{12}$ yd deep. Do they need to know the area or the volume to know how much concrete is needed? Come up with a strategy to figure out how much concrete they should pour.

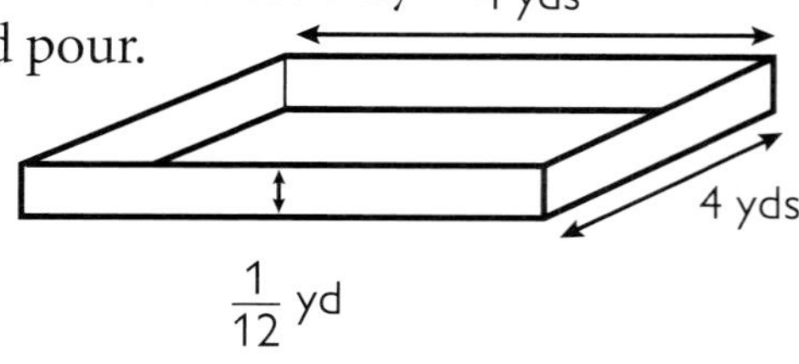

Making Connections

18. Imagine you have been hired by a famous clothing designer. It is your job to purchase the fabric for the upcoming designs. Your boss asks you to draw a smaller version of the designer's most successful scarf using a scale factor of 1 : 3. The original scarf is one yard long and one yard wide.

a. Draw a pattern with measurements for the new smaller scarf.

b. The company will make one hundred smaller scarves using your new pattern. How much fabric should you purchase?

Stepping into Gulliver's Worlds

Applying Skills

Complete the following chart by supplying the missing equivalents as decimals or fractions.

	Decimal	Fraction
1.	0.125	
2.		$\frac{3}{8}$
3.	0.75	
4.	0.67	
5.		$\frac{1}{2}$
6.	0.2	
7.		$\frac{1}{10}$
8.		$\frac{1}{20}$

9. Measure the sides of the square in inches. What is the:

a. perimeter? **b.** area?

10. Use the metric system to measure the height, width, and length of the cube below.

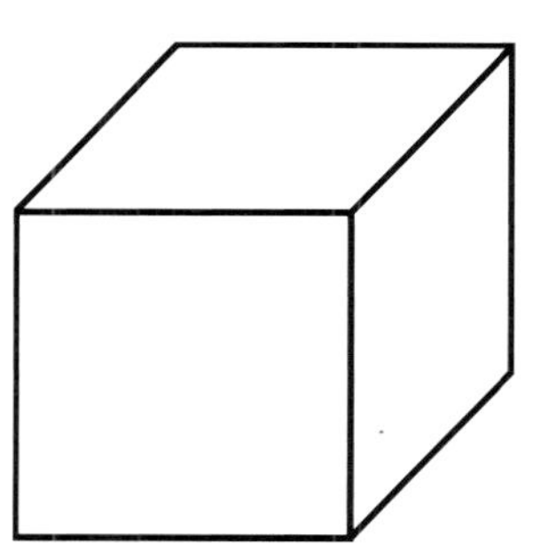

Extending Concepts

11. Shelley wants to cover a box with contact paper. The box is 1 ft high, 1 ft wide, and 1 ft deep.

a. Draw the box and show the measurements.

b. How many square feet of paper will she need to cover one face of the box?

c. How many square feet will she need to cover all sides of the box?

Making Connections

12. The creator of a famous theme park wanted children to feel bigger than life. The scale factor of objects in real life to objects in the park is 1:0.75.

a. Height of a street light?

Real life: 12 ft

Theme park:

12 feet

b. Length and width of a door?

Real life: 32″ × 80″

Theme park:

c. Height, width, and depth of a box?

Real life: 4 ft × 4 ft × 8 ft

Theme park:

STUDENT GALLERY

area
Scale

volume

Proportion

fraction

ratio

The Seeing and Thinking Mathematically project is based at Education Development Center, Inc. (EDC), Newton, MA, and was supported, in part, by the National Science Foundation Grant No. 9054677. Opinions expressed are those of the authors and not necessarily those of the National Science Foundation.

CREDITS: Photographs: Chris Conroy Photography • Beverley Harper (cover) • Duane Bibby: pp. 2–6, 8, 10, 12, 14–16, 18, 20, 22 • Saul Rosenbaum: pp. 26–28, 30, 32, 35–38, 42, 44–45.

Two Prudential Plaza, Suite 1175
Chicago, IL 60601

Printed in the United States of America

0-7622-0213-0

3 4 5 6 7 8 9 10 . 02 01 00 99